Federal Transit Administration

SECURITY AND EMERGENCY MANAGEMENT TECHNICAL ASSISTANCE FOR THE TOP 50 TRANSIT AGENCIES

FINAL REPORT

APRIL 2007

REPORT DOCUMENTATION PAGE		*Form Approved* *OMB No. 0704-0188*

Public reporting burden for this collection of information is estimated to average 1 hour per response, including the time for reviewing instructions, searching existing data sources, gathering and maintaining the data needed, and completing and reviewing the collection of information. Send comments regarding this burden estimate or any other aspect of this collection of information, including suggestions for reducing this burden, to Washington Headquarters Services, Directorate for Information Operations and Reports, 1215 Jefferson Davis Highway, Suite 1204, Arlington, VA 22202-4302, and to the Office of Management and Budget, Paperwork Reduction Project (0704-0188), Washington, DC 20503.

1. AGENCY USE ONLY (Leave blank)	2. REPORT DATE April 2007	3. REPORT TYPE AND DATES COVERED Final Report April 2007
4. TITLE AND SUBTITLE Security and Emergency Management Technical Assistance for the Top 50 Transit Agencies		5. FUNDING NUMBERS VT56/DV727
6. AUTHOR(S) Nick Bahr*, Erin Gorrie*, Mark Zannoni* Kevin Chandler**		
7. PERFORMING ORGANIZATION NAME(S) AND ADDRESS(ES) U.S. Department of Transportation Federal Transit Administration Office of Program Management Office of Safety and Security Washington, DC 20590		8. PERFORMING ORGANIZATION REPORT NUMBER DOT-VNTSC-FTA-07-01
9. SPONSORING/MONITORING AGENCY NAME(S) AND ADDRESS(ES) U.S. Department of Transportation Federal Transit Administration Office of Program Management Office of Safety and Security Washington, DC 20590		10. SPONSORING/MONITORING AGENCY REPORT NUMBER FTA-MA-90-5012-07
11. SUPPLEMENTARY NOTES * Booz Allen Hamilton ** Battelle Memorial Institute		
12a. DISTRIBUTION/AVAILABILITY STATEMENT This document is also available to the public through the National Technical Information Service, Springfield, VA 22161.		12b. DISTRIBUTION CODE

13. ABSTRACT (Maximum 200 words)

Between May 2002 and July 2006, the Federal Transit Administration (FTA) provided technical assistance to the top 50 transit agencies through the Security and Emergency Management Technical Assistance Program (SEMTAP). The scope and purpose of the program were:

- Review the transit agency's environment for security and emergency management
- Review, analyze, and make recommendations on security documents
- Develop methods to enhance security and emergency management procedures and training
- Develop and refine counter-terrorism tools
- Assess training needs and provide technical assistance for training
- Develop materials for security briefings and awareness
- Provide technical assistance for emergency tabletop exercises and planning for actual drills
- Provide guidance on how to conduct threat and vulnerability assessments (TVAs)

This report includes a program background and summary, the methodology used, findings and results gathered during the technical assistance visits and a description of the next generation technical assistance program.

14. SUBJECT TERMS Transit Security, Technical Assistance, Threat and Vulnerability Assessments, Counter-terrorism, Emergency Management, Preparedness			15. NUMBER OF PAGES 28
			16. PRICE CODE
17. SECURITY CLASSIFICATION OF REPORT Unclassified	18. SECURITY CLASSIFICATION OF THIS PAGE Unclassified	19. SECURITY CLASSIFICATION OF ABSTRACT Unclassified	20. LIMITATION OF ABSTRACT Unlimited

Acknowledgments

The Federal Transit Administration's Security and Emergency Management Technical Assistance Program for the Top 50 Transit Agencies was performed by Booz Allen Hamilton and Battelle (with the support of Transportation Resource Associates and Total Security Services International).

Booz Allen Hamilton and Battelle would like to extend full appreciation to the Federal Transit Administration and the following individuals who were instrumental in initiating this project and bringing it to a successful conclusion:

Harry Saporta
Director, FTA Office of Safety and Security (ex officio)

Michael Taborn
Director, FTA Office of Safety and Security

Richard L. Gerhart
FTA Team Leader, Security and Project Manager

In addition, FTA, Booz Allen Hamilton, and Battelle would like to offer special thanks to the General Managers/Chief Executive Officers and Security Directors/Police Chiefs of the 50 transit agencies for their industry leadership and willingness to participate in this voluntary program.

Disclaimer

This document is disseminated under the sponsorship of the Department of Transportation in the interest of information exchange. The United States Government assumes no liability for its contents or use thereof.

Table of Contents

Acknowledgments i
List of Exhibits iv
Chapter 1: Introduction 1
   I. Project Background 1
   II. Program Summary 2
Chapter 2: Methodology 5
Chapter 3: Findings and Results 9
   I. Observations 9
   II. Lessons Learned 10
   III. Transit Industry Products 12
Chapter 4: Transition to FTA's Next-Generation Security and Emergency
              Management Technical Assistance Program 15
   I. Program Overview 15
   II. Guiding Principles 16
   III. Gap Products 17

Appendix A. FTA Top 20 Security and Emergency Management Action
               Items List 19
Appendix B. Transit Agencies Participating in the FTA Top 50 SEMTAP 21

List of Exhibits

Exhibit 1. Locations of Transit Agencies Receiving FTA Readiness Assessments 1

Exhibit 2. Locations of Participating Transit Agencies in SEMTAP 3

Exhibit 3. Threat and Vulnerability Assessment Process 6

Exhibit 4. Top 50 Security and Emergency Management Technical Assistance Process 7

Exhibit 5. Frequency of Technical Assistance Product Categories Requested by
   Transit Agencies 13

Exhibit 6. Security and Emergency Management Capability Maturity Model 15

Chapter 1: Introduction

I. Project Background

Transit is a critical, high-risk, and high-consequence national asset. Each day, public transportation systems provide mobility to millions of Americans and serve the largest economic and financial centers in the nation. Transit systems are designed to provide open, easy access for passengers and to operate alongside or near our largest business and government buildings, intermodal transportation centers, and many of our nation's most visible icons. These facts make transit a prime target for terrorist attack.

Shortly after the terrorist attacks of September 11, 2001, the Federal Transit Administration (FTA) conducted security readiness assessments at the largest transit properties across the nation. These assessments were performed to provide an industry-wide understanding of the current level of security and emergency preparedness of the U.S. transit industry. Exhibit 1 depicts the 21 metropolitan areas representing the 37 transit agencies conducting readiness assessments.

Exhibit 1. Locations of Transit Agencies Receiving FTA Readiness Assessments
(Some locations reflect multiple transit agencies)

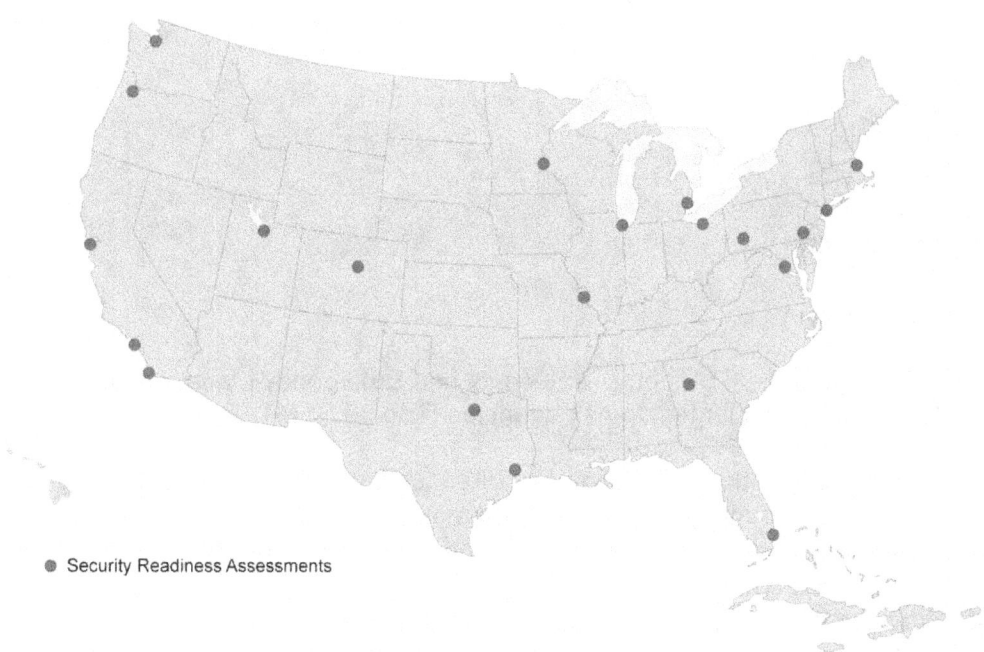

● Security Readiness Assessments

The findings of the readiness assessments led to the development and implementation of a set of FTA security and emergency management initiatives, including:

- Distributing FTA security and emergency management toolkits to transit agencies
- Creating additional transit-based security and emergency management resource documents, such as *The Public Transportation System Security and Emergency Preparedness Planning Guide*

- Providing grants of up to $50,000 to the top 100 transit agencies to conduct drills and exercises
- Developing new security and emergency management training courses, videos, and employee pocket guides through the National Transit Institute and the Transportation Safety Institute
- Creating the Connecting Communities Regional Inter-Agency Forums two-day workshop and delivering it to 18 regions across the country
- Conducting research and development of technology-based tools such as the PROTECT chemical sensor project
- Convening Security Roundtables that bring together security chiefs from the top 50 transit agencies for peer-to-peer informational exchanges

In addition to these security initiatives, FTA offered a voluntary on-site Security and Emergency Management Technical Assistance Program to the 50 largest transit agencies. Given the wide range of size, operating environment, and relative levels of security preparedness among the 50 largest transit agencies, the FTA on-site technical assistance program needed to allow flexibility, yet still provide a consistent structured systems approach. To meet these challenging project goals, and ultimately to serve as a benchmark assessment tool for the entire transit industry, FTA developed its Top 20 Security Action Items Checklist through which transit agencies could assess their current state of security readiness, identify any gaps, and improve their security posture. The Top 20 Security Items were organized into eight categories (see Appendix A for the complete list):

- Management and accountability
- Security problem identification
- Employee selection
- Training
- Audits and drills
- Document control
- Access control
- Homeland security

II. Program Summary

The purpose and scope of the FTA's on-site Security and Emergency Management Technical Assistance Program for the 50 largest transit agencies (Top 50 SEMTAP) were to:

- Review the transit agencies' environment for security and emergency management
- Review, analyze, and make recommendations on security documents
- Develop methods to enhance security and emergency management procedures and training
- Develop and refine counterterrorism tools
- Assess training needs and provide technical assistance for training
- Develop materials for security briefings and awareness
- Provide technical assistance for emergency tabletop exercises and planning for actual drills
- Provide guidance on how to conduct threat and vulnerability assessments (TVAs)

Top 50 SEMTAP began in May 2002 and spanned nearly four years, concluding in July 2006. The program provided technical assistance to the 50 largest transit agencies across the United States in metropolitan areas shown in Exhibit 2. (For a list of the 50 specific transit agencies, see Appendix B.)

Exhibit 2. Locations of Participating Transit Agencies in SEMTAP
(Some locations reflect multiple transit agencies)

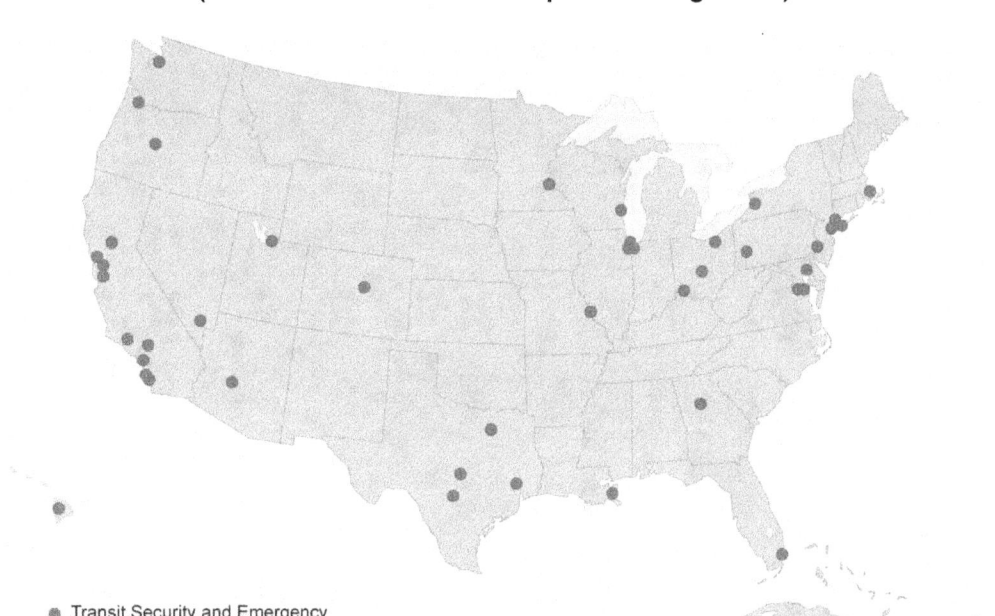

• Transit Security and Emergency Planning and Technical Assistance

FTA contracted with two firms to assist in providing the technical assistance; each firm was assigned 25 specific transit agencies. To address regional coordination, transit agency assignments by contractor were selected based as much as possible on a regional consideration; for example, one contractor was assigned all participating transit agencies in the San Francisco Bay area, while the other contractor was assigned all transit agencies in the Los Angeles area.

This Top 50 SEMTAP Final Report includes:

- A summary of the methodology used for Top 50 SEMTAP
- Key findings and observations from the program
- A discussion of FTA's next-generation SEMTAP activities

Please note that this Final Report is intended to be a summary document and does not include any specific sensitive security information (SSI) gathered and developed during the Top 50 SEMTAP process.

Chapter 2: Methodology

Top 50 SEMTAP followed a multistep process at each transit agency. The methodology that was created ensured a consistent process at each of the 50 transit agencies while still providing flexibility to address each transit agency's uniqueness in terms of operating environment and specific technical assistance needs as requested by the transit agency.

The technical assistance at each property began with the signing of a Memorandum of Understanding (MOU) between FTA and the transit agency. While participation in Top 50 SEMTAP by the transit agencies was voluntary, a need existed to create and execute an MOU to document agreement on:

- The scope of Top 50 SEMTAP
- Proper handling of SSI during the process
- The process for sharing the final products outside FTA and the transit agency
- The follow-up implementation plan for the transit agency

Once the MOU was signed, the next step in the process was an advanced planning meeting (APM), whereby the transit agency's security director and other members of the management met with FTA's project manager and contractor. The purposes of the APM were to:

- Brief the transit agency on the overall Top 50 SEMTAP process
- Discuss the set of standard FTA Top 50 SEMTAP deliverables: a threat and vulnerability assessment (TVA), a Top 20 Security Action Items List assessment, and a follow-up implementation plan
- Discuss additional possible deliverables based on the transit agency's specific needs and requests
- Schedule the remainder of the technical assistance effort, including when the contractor team would be working on site at the transit agency

Two-thirds of the 50 transit agencies receiving technical assistance had already received a TVA, determining the agency's ability to deter, detect, respond to, and recover from terrorism. For these transit agencies, the Top 50 SEMTAP contractor reviewed the status of the TVA recommendations and created a tracking matrix that addressed the status and disposition of the TVA recommendations.

For the 15 transit agencies participating in Top 50 SEMTAP that had not previously received a TVA from FTA, the first action undertaken by the Top 50 SEMTAP contractor was to perform a TVA.

The TVA methodology was used as a baseline standard. The Top 50 SEMTAP technical assistance teams reviewed the current state of security against previously identified threat scenarios, evaluated processes and existing countermeasures for mitigating threats, and identified proposed potential mitigation alternatives. Exhibit 3 illustrates the TVA methodology.

Exhibit 3. Threat and Vulnerability Assessment Process

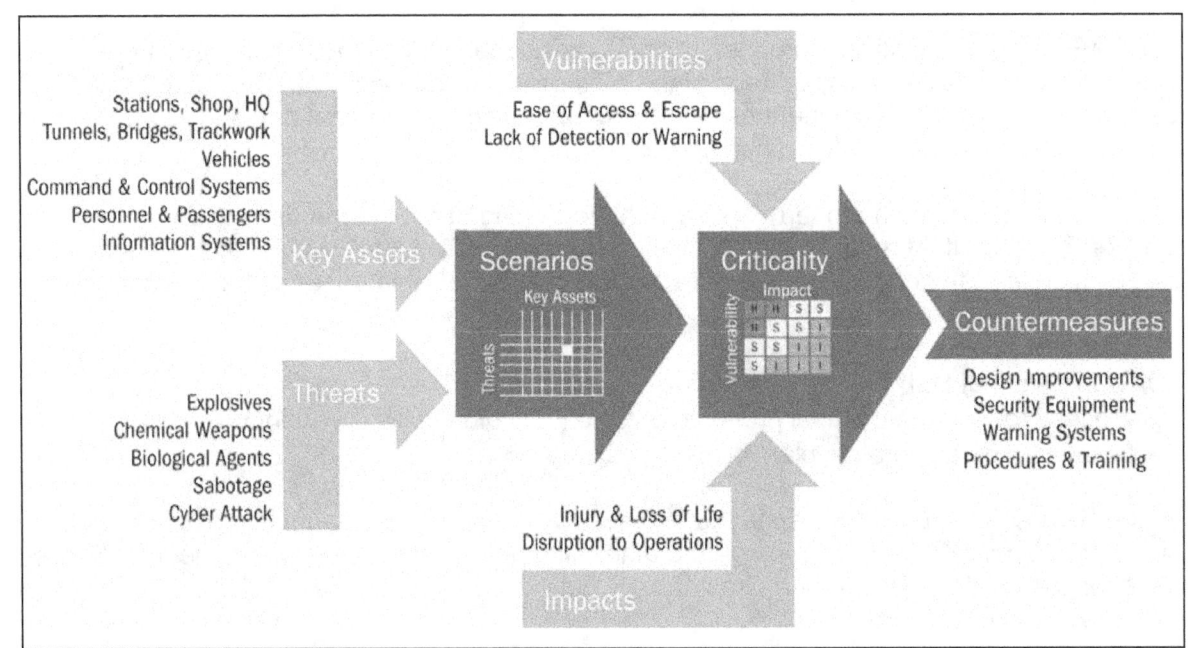

The TVA process included interviews with key agency personnel ranging from frontline staff to executive management, physical on-site inspections, and a review of security and emergency plans and procedures. From the TVA and the Top 20 Security Action Items List, a transit agency's security readiness and program gaps needing attention were established. Based on the approach of "filling the gaps," specific technical assistance deliverable products were determined. Most of the 50 transit agencies requested specific additional technical assistance in a range of areas. (See Chapter 3, Section III – Transit Industry Products for more information.)

Once the transit agency and FTA were in agreement on the specific products, the contractor's technical assistance team started developing the deliverables for the transit agency. Typically the contractor team needed a week on site at the transit agency for data-gathering activities such as interviews with key staff, site-specific inspections of key infrastructure, and document reviews.

When the draft set of deliverables was completed, the contractor technical team conducted a product review/delivery meeting with FTA and the transit agency that covered:

- A review of the draft deliverables
- FTA and transit agency agreement on the follow-up deliverables implementation plan
- An updated Top 20 Security Action Items List assessment based on the completed technical assistance

The last step in the process was for the contractor technical team to process any remaining edits and to prepare and distribute final reports/products (in both notebook and CD formats) to the transit agency and FTA. Exhibit 4 summarizes this process.

Exhibit 4. Top 50 Security and Emergency Management Technical Assistance Process

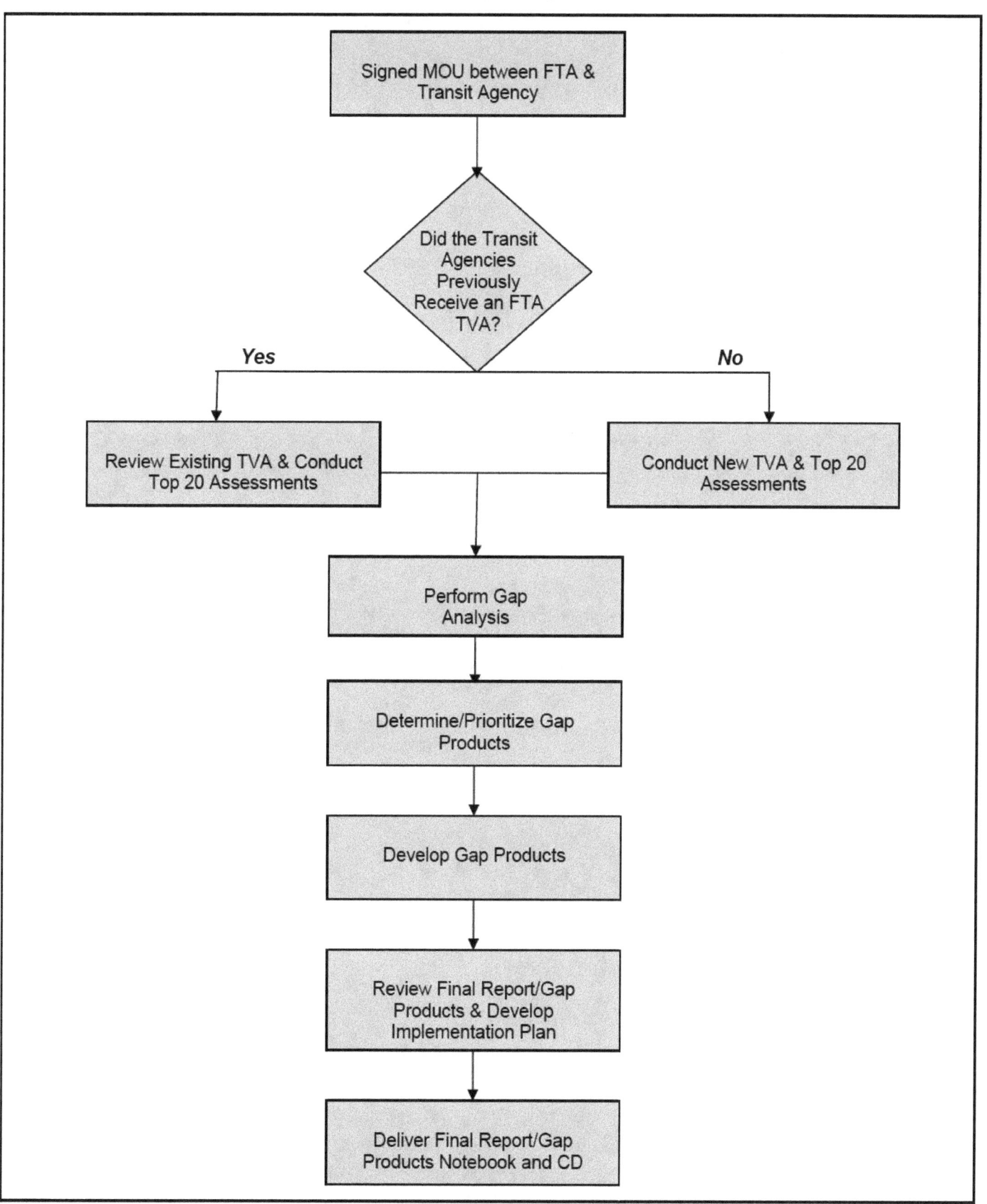

Chapter 3: Findings and Results

I. Observations

As the Top 20 Security Action Items List assessments and TVAs revealed, wide variation is evident in the levels of security preparedness among the 50 transit agencies studied. All 50 agencies exhibited significant strengths in critical areas of importance.

Areas needing improvement also existed, which, if left unattended, may leave these transit agencies vulnerable to terrorist or criminal activity. At some transit agencies, security is still oriented toward reacting to external factors and is not a core focus of management. Some of the smaller transit agencies within the Top 50 list consider themselves to be unlikely targets for terrorist attacks. (See Chapter 4, Section I – Program Overview for more discussion of the transit maturity model.)

Security Environment – Throughout the course of their on-site work, the contractor technical teams noted the security environments of the transit agencies, such as high-risk targets on or near transit routes. These targets included major tourist attractions, government buildings, chemical and nuclear power plants, railroad yards, and military bases, all of which have their own specific security protection measures.

The contractor technical teams noted that local jurisdictions were aware of these high-risk targets and had to a large degree worked closely with emergency service responders and the transit agency to develop coordinated security and emergency management plans.

Perimeter Security – Perimeter security of transit facilities includes closed circuit television (CCTV) systems, lighting, perimeter access points, and physical barriers. Perimeter security in general is an area where improvements are required, though the level of security varied greatly among transit agencies.

CCTV – CCTV systems were found to be in use at most properties. However, the effectiveness of these systems varied considerably. Some CCTV systems were either outdated or not systematically designed or deployed while other agencies utilized state-of-the-art systems at their facilities and on their revenue service vehicles.

Entry and Exit Security, and Access Control – Entry and exit security refers to measures for controlling the flow of people into and out of a transit facility as well as the delivery of packages and materials. Again, the level of effectiveness varied from property to property as some agencies were improving their methods and technologies. Some transit agencies had limited internal access control measures, while others had sophisticated systems.

Employee Identification Programs – Most properties had an employee identification program consisting of agency-issued photo identification badges. However, employee compliance with wearing badges and management oversight of this compliance varied among transit agencies.

Security Planning – Security planning encompasses the plans, policies, and procedures that govern the scope, operations, and management of security personnel, activities, and technologies. Throughout the technical assistance program, security planning at transit agencies was being continually improved or enhanced, reflecting the important role that planning plays as the foundation of a good security program.

Coordination with Local Emergency Agencies – Excellent liaison with first responders was found at a majority of the 50 transit agencies. Good coordination with first responders occurred in responding to both regional emergencies and specific transit-related emergencies. This coordination included conducting regularly scheduled tabletop and functional exercises that addressed topics such as hostage situations on buses and discovery of suspicious powders and other substances. Numerous instances of ongoing, active training took place between law enforcement SWAT teams and transit agencies. Many agencies had existing programs that allowed SWAT teams to train on agency buses, and at least one agency had a program to donate retired buses to police departments for training and familiarization.

Security Awareness – The level and amount of security awareness training throughout the transit industry varied from property to property. For agencies with less than optimal levels of training across their staff, the cost of such training was cited.

Strengths – Overall, transit agencies have made significant strides in their security awareness and readiness in the post-9/11 timeframe. Proactive agencies have unilaterally undertaken many initiatives. Dissemination of industry-recommended practices helps transit agencies with limited resources or minimal experience to implement programs that might otherwise be inaccessible or unavailable. The contractor technical teams noted numerous strengths common among the 50 participating transit agencies, summarized below:

- Many transit agencies now have dedicated, high-visibility security and emergency management managers who are striving to ensure that their security and emergency management documentation is complete, approved, and incorporated into training programs. The security managers generally participate in regional counterterrorism councils, and they have the full support of upper management.
- All agencies interviewed considered the sharing of security information with their peers as a key strategy to spread best practices and information.
- Transit agencies have significant experience in responding to natural disasters and accidents in their local and regional areas. This experience has generally been used as a foundation for expanding and enhancing security and emergency management programs.
- Transit agencies are committed to updating critical security documents such as System Security and Emergency Preparedness Plans. Many did this through or concurrently with the Top 50 SEMTAP.
- Many transit agencies are upgrading their CCTV systems and their perimeter and access control systems.
- The transit industry has widely embraced FTA's Transit Watch program to enhance public awareness participation in the form of "eyes and ears" programs.
- Most transit agencies perform thorough pre-employment background checks.
- Safety and security drills are regularly held at many agencies, often in partnership with local emergency response agencies.

II. Lessons Learned

Numerous lessons were learned during the Top 50 SEMTAP:

Information Overload – Transit agency security and emergency management professionals receive federal, state, and local guidance documents, bulletins, and correspondence related to security and emergency management. The challenge for transit agency personnel is to digest this large volume of information and translate it into meaningful transit-specific activities and guidance. Several transit agencies requested contractor technical team help in analyzing existing and emerging guidance and providing recommendations.

A tremendous amount of guidance material is provided to the transit industry by the federal government, but many of the documents are outdated, do not address counterterrorism, or are not directly applicable to transit operations because they were originally written for other industries or modes of transportation.

Comprehensive Sustainable Security Training Programs – Security and emergency management training often does not reach the right audience. It is either too generic or only targets a very small subset of the transit agency workforce. Much of the training material is disjointed and unrelated and is thus confusing to staff.

The transit industry has struggled with establishing and sustaining comprehensive security and emergency management training programs for its employees. Most transit agencies can afford to provide security training only for new employees as part of their orientation activities.

Some rail transit employees may receive additional refresher security training as a module added into the required annual recertification training for rail operators. However, the significant costs associated with providing security training for bus and rail operators and other frontline transit employees (due to the need to backfill these positions during training or to pay overtime rates) have stymied transit agencies in being able to provide needed ongoing security and emergency management training.

Intelligence – A considerable amount of security-related information is being provided to transit agencies as intelligence. This large volume of information may or may not involve the transit industry or the specific transit agency. Both specificity and timeliness have been identified by transit agencies as areas needing intelligence improvements. A few of the largest transit agencies have resorted to creating their own internal intelligence/information units to address current transit industry intelligence program deficiencies.

Technology – Transit agencies need introduction of appropriate technology that is:

- Proven and reliable in the transit environment, which often includes harsh physical conditions
- Justifiable from a TVA cost-benefit analysis weighing the risks, effectiveness of countermeasures, and cost
- Appropriate for the transit environment, balancing staff availability, budget, and skill sets

In recent years, the transit industry has experienced a substantial introduction of technology in areas such as automated systems and advanced communications, through such initiatives as Intelligent Transportation Systems (ITS). The recent focus on security and emergency management has introduced many actual and potential technological solutions. The challenge for the transit industry has been learning about these technologies and how to use them, appropriately applying them to specific transit agencies' needs, and accurately planning for all associated life-cycle costs, not just the capital costs of procuring the technology (i.e., costs for installation, operations, maintenance, and product updates). One of these needs is correlating performance measures of the technology with cost of the investment. Guidance in this area is limited, particularly as new technologies and products are introduced frequently into the market.

Designing in Security – Transit agencies are often unsure how to incorporate security into the design of new transit infrastructure projects and as a result end up "bolting on" security measures after the infrastructure is constructed. New procurements and capital projects, including revenue service vehicles, passenger terminals and stations, and other related infrastructure, need to take into account security and emergency management while the project is in the design stage.

Emergency Management – Emergency management encompasses both emergency and preparedness activities associated with planning before an emergency occurs, such as regional planning and coordination, internal transit agency preparedness, and passenger awareness. Regional coordination focuses on areas such as drills and exercises, interoperable communications, and contingency plans. Drills and exercises should be designed with clear objectives and should focus on testing plans, procedures, and equipment.

Concerns include whether evaluations of a drill or tabletop exercise (such as after-action reports) are completely objective and accurately document observed weaknesses. (At a minimum, evaluation should be clearly focused on measuring emergency management and communications.) Another issue is making sure that problems discovered during drills or exercises have a clear path to resolution and documentation. The drill or exercise must be credible and well executed to maximize the value of the activity.

Insufficient understanding of Incident Command Systems (ICS) and National Incident Management Systems (NIMS) is prevalent in the transit industry. More effort is needed to properly train the appropriate transit agency staff, especially those in positions of operations supervision (such as communications/control center personnel, street/field supervisors) and security/law enforcement personnel on ICS and NIMS. The goal of this training is for staff to understand both their role in an emergency and the emergency responders' expectations.

Significant attention has been given to the need for communications plans, appropriate and compatible equipment, and backup equipment for communications and dispatching. Interoperability of communications equipment in a given region and backup communications have been identified as needs; however, the cost of appropriate solutions is typically extremely high. One of the continuing needs in the transit industry is for a plan of operations during a communications failure, that is, how operations would proceed when communications are interrupted or unavailable.

III. Transit Industry Products

During the multiyear SEMTAP effort, FTA developed a variety of product deliverables across the 50 transit agencies. Exhibit 5 presents the set of SEMTAP transit industry deliverables that were most frequently requested among the 50 agencies. By far the largest categories of requested products were in emergency management, security awareness and training, and Homeland Security Advisory System (HSAS) threat-level procedures. Though these were the largest categories, a broad range of products was developed. The Other category represents products developed at only a handful of transit agencies (such as guidance on access control or technology reviews) or at only one transit agency (such as an asset-specific TVA or development of an IT security policy).

Exhibit 5. Frequency of Technical Assistance Product Categories Requested by Transit Agencies

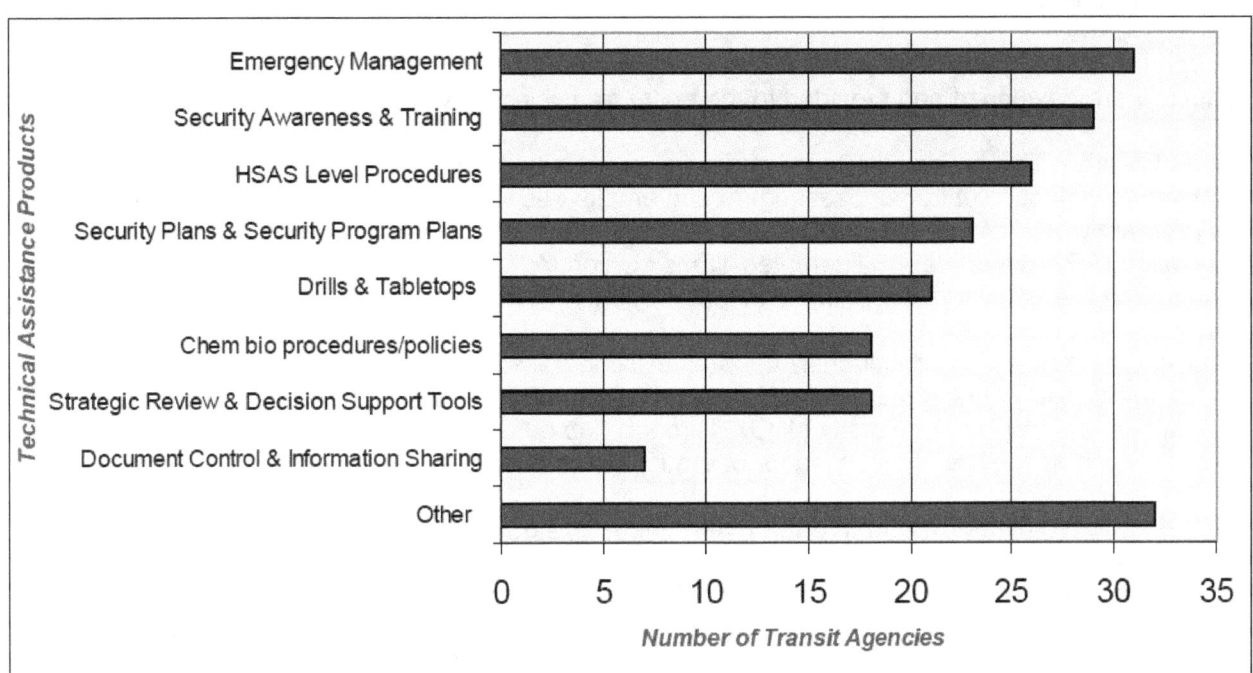

A representative list of the variety of Top 50 SEMTAP products developed is as follows:

- System security and emergency preparedness plans
- Business recovery plans
- Emergency response procedures, including chemical, biological, and radiological responses
- Security/police force models
- Drills and exercise support/guidelines
- Disaster recovery plans
- Security funding support
- Threat and vulnerability assessments
- Security awareness programs
- Training support and programs
- Public awareness support
- Tunnel security and blast analyses
- Regional emergency coordination
- Threat-level response procedures/guidelines
- Vehicle accountability systems
- Evacuation plans
- Communications support/guidance
- Security in the procurement process guidelines

In addition to tracking the security readiness of the 50 largest transit agencies, FTA used the Top 20 Security Action Items matrix to identify industry gaps that might benefit from guidance documents. For example, the initial results for Top 20 Action Item 14 regarding Public Awareness Programs indicated that very few of the 50 largest transit agencies had developed and implemented an awareness program. As such, FTA, along with other industry partners, developed the Transit Watch program as a means to fill this gap.

Through the ongoing Top 20 matrix tracking process during Top 50 SEMTAP, the following gap products were developed and provided to the entire transit industry:

Transit Watch – This nationwide "eyes and ears" public awareness outreach campaign encourages the active participation of transit passengers and employees in maintaining a safe transit environment. This program was originally launched in 2003 in partnership with the American Public Transportation Association (APTA) and the Amalgamated Transit Union (ATU) and was recently updated in collaboration with the Department of Homeland Security (DHS).

Training Courses – FTA continuously updates existing transit security courses and adds new courses through its course contractors: National Transit Institute (NTI), Transportation Safety Institute (TSI), and Johns Hopkins University (JHU). Through Top 50 SEMTAP, additional course needs were identified, such as job-specific training for operations control center personnel.

HSAS Protective Measures for Transit Agencies – This document provides an approach that integrates a transit agency's entire security and emergency management program with DHS HSAS threat conditions. For example, if the HSAS threat level is raised from yellow to orange, specific additional actions should be taken by each transit agency. The guidance document was recently updated by FTA to include a systematic all-hazards methodology.

Immediate Actions (IAs) for Transit Agencies – This guidance document is intended to help transit agencies improve the effectiveness of the reaction and response of their front-line employees to potential and actual life-threatening incidents.

Security Design Considerations – This document provides security design guidelines for the major assets of transit systems – buses, rail vehicles, stations, communications systems, yards, and depots – as well as a preliminary assessment of the vulnerabilities to various methods of attack inherent in each asset. It addresses access management, systems integration, and communications, discusses major threats, and introduces Crime Prevention Through Environmental Design (CPTED) in transit systems.

Chapter 4: Transition to FTA's Next-Generation Security and Emergency Management Technical Assistance Program

I. Program Overview

Security and emergency management programs at transit agencies are maturing. Transit agencies are embracing security and integrating it into their daily operations. In years past, transit agencies were largely reactive, with basic plans in place, but a focus on handling emergencies as they occurred.

As transit agencies' capabilities mature, they have become more proactive and holistic, incorporating security and emergency management activities and responsibilities across all departments, job categories, and infrastructure facilities, as well as through the various stages of an incident – from the threat of an event to an actual event to response and recovery. This vision of a capability maturity model for transit agencies is illustrated in Exhibit 6.

Exhibit 6. Security and Emergency Management Capability Maturity Model

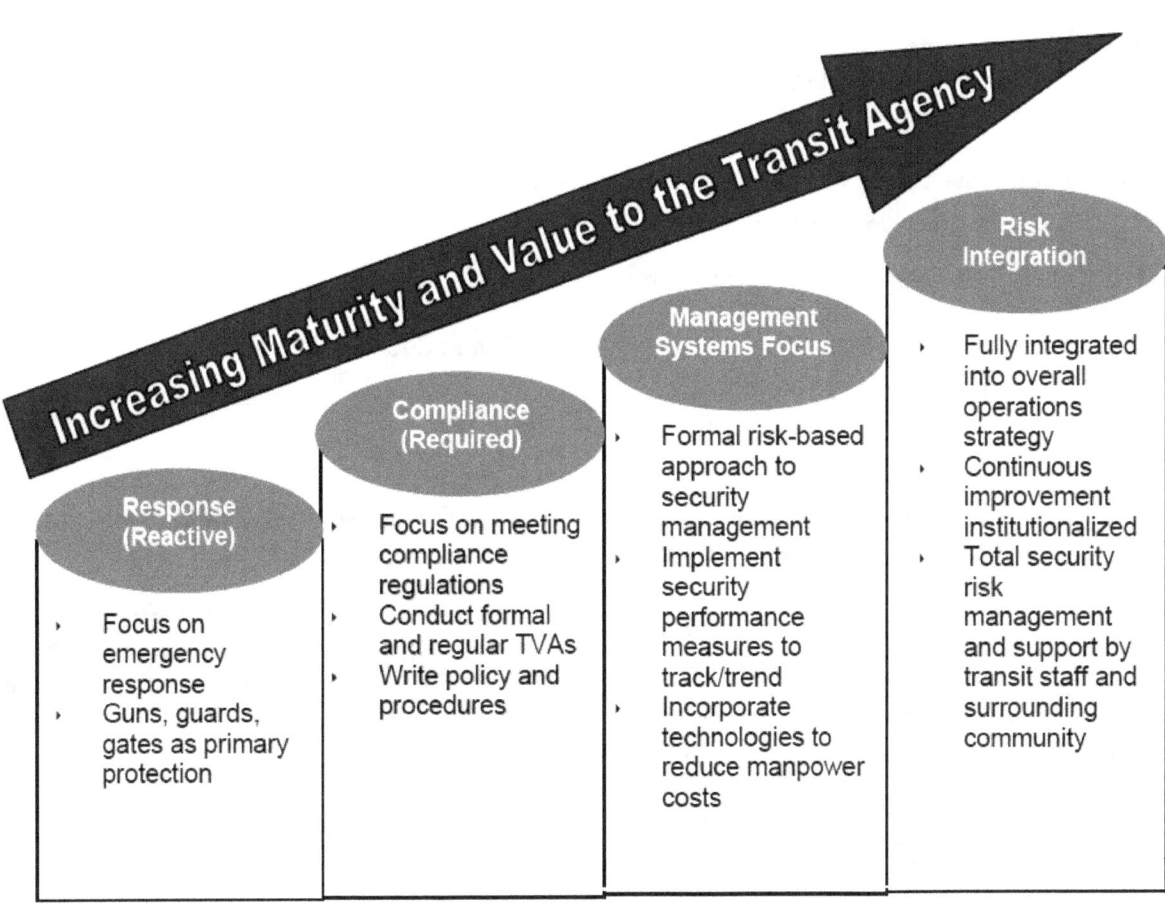

Transit agencies must continue to build and improve upon the substantial security and emergency management improvements accomplished in the post-9/11 environment. Continual improvement means that on a regular basis:

- Security and emergency management plans, programs, and protocols need to be reviewed and updated
- Employees need to be trained and retrained on the contents of the plans, programs, and protocols
- Employees need to be tested on how well they have been trained through frequent drills and exercises
- After-action reports and other critiques of drills and exercises need to identify any needed improvements to the plans, programs and protocols

The next generation of FTA's security and emergency management technical assistance must build on the experience and lessons learned from the original Top 50 SEMTAP activities. This next generation of technical assistance, started in September 2006, has a twofold scope:

- Develop a comprehensive, coordinated, and continuous security and emergency management planning process for FTA
    - product: FTA Five Year Security and Emergency Management Strategic Plan (to be updated annually)

- As part of the Strategic Plan, develop a strategic analysis process that identifies program gaps and develop products to fill the gaps
    - product: a set of industry gap products (to be updated annually)

The first step in the development of the next-generation technical assistance program is largely information gathering, with three sources of information:

1. The Top 50 SEMTAP Final Report (this report)
2. Interviews of key stakeholders
3. Reviews of critical documents (both completed/distributed documents and documents still under development)

II. Guiding Principles

In developing the process associated with creating the FTA NexGen SEMTAP Strategic Plan, three key guiding principles will serve as basic foundations:

(1) U.S. Department of Transportation/U.S. Department of Homeland Security Memorandum of Understanding Annex regarding public transportation security roles and responsibilities: this guiding agreement between FTA, Transportation Security Administration (TSA), and Grants & Training (G&T) establishes the inter-agency framework for the parties to collaborate on all matters pertaining to public transportation security.
- for NexGen SEMTAP this means: TSA and G&T will be requested to participate in the development and review of all NexGen SEMTAP products (Strategic Plans and gap products)

(2) FTA has benefited from close working relationships with key industry stakeholder organizations such as APTA, Community Transportation Association of America (CTAA),

Transportation Research Board (TRB), American Association of State Highway Officials (AASHTO) and the ATU
- for NexGen SEMTAP this means: APTA, CTAA, TRB, AASHTO and ATU will be requested to participate in the development and review of all NexGen SEMTAP products (Strategic Plans and gap products)

(3) FTA has developed three strategic security priorities – employee security training, public awareness, and emergency preparedness – and has worked closely with TSA and G&T to position these three priorities as industry principles.
- for NexGen SEMTAP this means: all NexGen SEMTAP products (Strategic Plans and gap products) developed will support these three strategic security priorities.

III. Gap Products

In annual updates of the Five Year Strategic Plan, new sets of gap products will be identified and prioritized through the strategic analysis process. To jumpstart this effort for the initial Strategic Plan cycle, a set of gap products that emerged from the Top 50 SEMTAP are under development:

- SSI guidance document – this product provides guidance for transit agencies in terms of proper designation, labeling, and handling of SSI.
- Resource links for the updated TSA/FTA Security and Emergency Management Action Items for Transit Agencies – FTA has collaborated with TSA to update the FTA's Top 20 Security Action Items List into the new TSA/FTA Security and Emergency Management Action Items for Transit Agencies. This gap product would provide links to additional resource documents and guidance materials for each action item.
- Security forces manpower planning model – this model was developed at two transit agencies as part of the Top 50 SEMTAP. This work would continue with additional model testing and validation at a larger cross section of transit agencies, with the intent of developing and distributing a generic, scalable security forces manpower planning model to the transit industry.
- Testing detailed protective measures implementation – the recently updated FTA "Transit Agency Security and Emergency Management Protective Measures" resource document includes an appendix that presents a suggested advanced systematic approach for transit agencies to consider when developing their protective measures plans, programs, and protocols. This effort would involve validation testing of this advanced approach at a cross section of transit agencies.

Appendix A. FTA Top 20 Security and Emergency Management Action Items List

Management and Accountability
1. Written security program and emergency management plans are established.
2. The security and emergency management plans are updated to reflect anti-terrorist measures and any current threat conditions.
3. The security and emergency management plans are an integrated system security program, including regional coordination with other agencies, security design criteria in procurements, and organizational charts for incident command and management systems.
4. The security and emergency management plans are signed, endorsed, and approved by top management.
5. The security and emergency management programs are assigned to a senior level manager.
6. Security responsibilities are defined and delegated from management through to the front-line employees.
7. All operations and maintenance supervisors, forepersons, and managers are held accountable for security and emergency management issues under their control.

Security Problem Identification
8. A threat and vulnerability assessment resolution process is established and used.
9. Security sensitive intelligence information sharing is improved by joining the FBI Joint Terrorism Task Force (JTTF) or other regional anti-terrorism task force; the Surface Transportation Intelligence Sharing and Analysis Center (ST-ISAC); and security information is reported through the National Transit Database (NTD).

Employee Selection
10. Background investigations are conducted on all new front-line operations and maintenance employees.
11. Criteria for background investigations are established.

Training
12. Security orientation or awareness materials are provided to all front-line employees.
13. Ongoing training programs on safety, security, and emergency procedures by work area are provided.
14. Public awareness materials are developed and distributed on a system-wide basis.

Audits and Drills
15. Periodic audits of security and emergency management policies and procedures are conducted.
16. Tabletop and functional drills are conducted at least once every six months and full-scale exercises, coordinated with regional emergency response providers, are performed at least annually.

Document Control
17. Access to documents of security critical systems and facilities are controlled.
18. Access to security sensitive documents is controlled.

Access Controls for Contractors and Visitors
19. Background investigations are conducted of contractors or others who require access to security critical facilities, and ID badges are used for all visitors, employees, and contractors to control access to key facilities.

Homeland Security
20. Protocols have been established to respond to the Office of Homeland Security Threat Advisory Levels.

* The FTA Top 20 was updated to become TSA/FTA Security and Emergency Management Action Items for Transit Agencies.

Appendix B. Transit Agencies Participating in the FTA Top 50 SEMTAP

Rank*	City/State	Transit Agency	Completion Date	Contractor Team
1	New York, NY	New York City Transit Authority (NYCTA)**	4/2004	Booz Allen Hamilton
2	Chicago, IL	Chicago Transit Authority (CTA)	8/2005	Battelle
3	Los Angeles, CA	Los Angeles County Metropolitan Transportation Authority (LACMTA)	5/2005	Booz Allen Hamilton
4	Boston, MA	Massachusetts Bay Transportation Authority (MBTA)	6/2005	Battelle
5	Washington, DC	Washington Metropolitan Area Transit Authority (WMATA)	1/2004	Battelle
6	Philadelphia, PA	Southeastern Pennsylvania Transportation Authority (SEPTA)	7/2006	Battelle
7	San Francisco, CA	San Francisco Municipal Railway (MUNI)	3/2004	Battelle
8	Newark, NJ	New Jersey Transit (NJT)	4/2004	Battelle
9	Atlanta, GA	Metropolitan Atlanta Rapid Transit Authority (MARTA)	5/2004	Battelle
10	New York, NY	New York City Department of Transportation (NYCDOT)	12/2005	Booz Allen Hamilton
11	Baltimore, MD	Maryland Transit Administration (MTA)	5/2005	Booz Allen Hamilton
12	New York, NY	Long Island Rail Road (LIRR)**	4/2004	Booz Allen Hamilton
13	Seattle, WA	King County Metro Transit (KCMetro)	3/2004	Battelle
14	Oakland, CA	San Francisco Bay Area Rapid Transit District (BART)	10/2004	Battelle
15	Houston, TX	Metropolitan Transit Authority of Harris County (Metro)	6/2005	Battelle
16	Portland, OR	Tri-County Metropolitan Transportation District of Oregon (TriMet)	9/2005	Battelle
17	Miami, FL	Miami-Dade Transit (MDT)	10/2005	Battelle
19	Denver, CO	Denver Regional Transit District (RTD)	3/2005	Battelle
20	Pittsburgh, PA	Port Authority of Allegheny County (PAAC)	1/2006	Battelle
21	Minneapolis, MN	Metro Transit	5/2005	Battelle
22	Chicago, IL	Metra	12/2004	Battelle
23	New York, NY	Metro-North Railroad (MNRR)**	4/2004	Booz Allen Hamilton
24	Milwaukee, WI	Milwaukee County Transit System (MCTS)	3/2006	Booz Allen Hamilton
25	Oakland, CA	Alameda Contra-Costa Transit District (AC Transit)	5/2004	Battelle

Appendix B. Transit Agencies Participating in the FTA Top 50 SEMTAP (cont.)

Rank*	City/State	Transit Agency	Completion Date	Contractor Team
26	Honolulu, HI	Honolulu Public Transit Department (TheBus)	5/2006	Booz Allen Hamilton
27	Cleveland, OH	Greater Cleveland Regional Transit Authority (GCRTA)	8/2004	Battelle
28	Dallas, TX	Dallas Area Rapid Transit (DART)	5/2006	Booz Allen Hamilton
29	New Orleans, LA	New Orleans Regional Transit Authority (NORTA)	5/2006	Booz Allen Hamilton
30	Orange County, CA	Orange County Transportation Authority (OCTA)	12/2004	Booz Allen Hamilton
31	San Jose, CA	Santa Clara Valley Transportation Authority (VTA)	1//2003	Battelle
32	Saint Louis, MO	METRO	1/2006	Battelle
33	Las Vegas, NV	Regional Transportation Commission of Southern Nevada (RTC)	12/2005	Booz Allen Hamilton
34	San Antonio, TX	VIA Metropolitan Transit	5/2006	Booz Allen Hamilton
35	San Juan, PR	Tren Urbano	3/2006	Booz Allen Hamilton
36	Detroit, MI	Detroit Department of Transportation (DDOT)	1/2005	Booz Allen Hamilton
37	San Diego, CA	San Diego Metropolitan Transit System (MTS)	4/2006	Battelle
38	Austin, TX	Capital Metro	5/2006	Booz Allen Hamilton
39	Arlington Heights, IL	Pace Suburban Bus	11/2004	Battelle
40	Phoenix, AZ	Phoenix Public Transit Department (PTD)	3/2006	Booz Allen Hamilton
41	New York, NY	Long Island Bus (LIB)**	4/2004	Booz Allen Hamilton
42	Buffalo, NY	Niagara Frontier Transit Authority (NFTA)	12/2005	Booz Allen Hamilton
45	Sacramento, CA	Sacramento Regional Transit District (SacRTD)	3/2006	Battelle
46	San Juan, PR	Autoridad Metropolitana de Autobuses (AMA)	3/2006	Booz Allen Hamilton
47	Cincinnati, OH	Southwest Ohio Regional Transit Authority (SORTA)	7/2006	Battelle
48	Long Beach, CA	Long Beach Transit (LBT)	5/2006	Booz Allen Hamilton
49	Salt Lake City, UT	Utah Transit Authority, (UTA)	5/2006	Booz Allen Hamilton
50	Santa Monica, CA	Santa Monica Transit, Big Blue Bus (BBB)	5/2006	Booz Allen Hamilton
53	Hampton, VA	Hampton Roads Transit (HRT)	8/2002	Booz Allen Hamilton
54	Columbus, OH	Central Ohio Transit Authority (COTA)	3/2006	Battelle
60	Seattle, WA	Washington State Ferry System (WSF)	4/2005	Booz Allen Hamilton

* Ranking is based on FY2000 National Transit Database annual ridership information.
** Technical assistance provided to the four New York City transit operating agencies was combined into one overall effort, under the NY MTA.

www.ingramcontent.com/pod-product-compliance
Lightning Source LLC
Chambersburg PA
CBHW081818170526
45167CB00008B/3454